Health Informatics for Clinicians
© Copyright 2012 Shane O'Hanlon

I0397541

Introduction

If you've ever hear people mention Health informatics (or healthcare informatics, or medical informatics or eHealth) and weren't quite sure what they were talking about – this book is for you.

If you are aware that there is a wide-ranging move towards using computers in more and more aspects of clinical practice, and feel like you are being left behind – this book is for you.

If you are a health professional who cares about your patients, and want to stay abreast with recent developments, this book is for you.

If you already know these things, you are not likely to get much more out of this! I've written this book as a primer or introduction to the area, and you might find it overly basic. But if you could do with brushing up, it might be useful.

I've really enjoyed putting this material together, and I hope that you enjoy reading it. It is not overly technical (I hope) and should be accessible to anyone, of any age.

I'd really appreciate any comments you have, whether you like or hate it. This will make future editions stronger. Once you finish, let me know if you would like to read more, and I will oblige!

Dr Shane O'Hanlon
Limerick, Ireland
December 2012

How to use this book:

The text is divided into chapters which each have an individual focus, but should flow from one to the next. You will get most out of it if you read it from start to finish. However you should feel free to skip along if you are already comfortable with any particular area. Each chapter ends with summary points, so you can get a quick idea of what each chapter is about by reading those.

You don't need a computer to get through it, but I have included some web links - so you can click through whenever you like to explore the recommended online content.

Index

Chapter 1: What is Health Informatics?

Before we get into anything here, it is useful to see what comes into your own head when we mention Health Informatics.

Take 10 seconds to come up with some words that come to mind when you think about Health Informatics.

Computers?

When I do this exercise with my class, the most common perception of Health Informatics is that it's mainly about computers. Many of you will probably also agree with this. But is it correct?

At a conference, Barnett and Sukenik noted that there was a large impending crisis in healthcare, particularly with increasing demand for service, and a lack of personnel. They argued that computer technology was the solution, apparently suggesting that it could fix many of the problems in healthcare. But they criticised the lack of commitment from the medical academic community and the health services.

Few people would disagree with the sentiment, but what's amazing about this is that the conference was in 1969!

(Quoted from G. Octo Barnett, H.J. Sukenik (1969), p. 268)

Even more than 40 years later, computers clearly have not lived up to our dreams and expectations. Why is this? Worse still is the fact that many of us can immediately think of disasters involving computers – for example:
- the Year 2000 (Y2K) bug,
- your records being lost because a system wasn't backed up
- bank websites going down just when you need them
- not to mention the fact that they can easily make a complete mess of your tables and formatting when making documents!

These examples are clearly of varying severity, but nobody actually got hurt by them. However, when computers are used in provision of healthcare, the result can actually be patients being harmed. Consider these two short case histories:

Case history: Therac-25 Linear Accelerators

The Therac-25 was a radiation therapy machine used in the 1980s. It was involved in at least six accidents between 1985-87. Patients were given massive doses of radiation – approximately 100 times the intended dose. One patient described getting an intense electric shock, and ran out of the room screaming. 3 of the 6 patients died as a result of the overdoses.

What happened?
At times, the system halted and gave an error code. Staff did not have a list of what the codes meant, so they pressed 'override' and went ahead with the treatment. The machine actually gave a low radiation reading so people were given multiple doses to 'compensate' for this.

Analysis:
Human error was the main contributor. The machine was programmed incorrectly, the operators went ahead despite the error code, and the manufacturer didn't believe the complaints that something was wrong. It seems the computer was not at fault...

en.wikipedia.org/wiki/Therac-25

(is Wikipedia an appropriate source? more on this later)

Case history: Missing X-rays
A child attending hospital did not have a broken bone detected when she first attended emergency. When she returned, she waited three-and-a-half hours before she was told her X-rays had been lost and more were being reprinted.

The patient then waited another hour, only to be told her details had been deleted off the hospital system. Her X-rays had to be taken again.

Analysis:
Again human error caused the problems here. The result is still excessive radiation exposure, though to a much smaller degree. But this story is from the very recent past - why is this still happening?

http://www.independent.ie/national-news/hospital-failed-to-detect-childrens-bone-fractures-1818900.html

A word about Technology:
In Health Informatics, technology is just a tool.

This is just the same as in a clinical field, such as Cardiology where a stethoscope is just a tool. It's not going to be much use, and can potentially cause harm, if the operator can't use it. As you can see from the above examples, simply using technology to do something does not make it safer. In fact it can make it more dangerous, expensive and time consuming.

In order to use technology effectively, the **information** on how to do this must be **communicated** effectively to the user.

So we could say, Health Informatics = ICT
I – Information
C – Communication
T – Technology

ICT is a commonly heard expression and it's a good way to remember what Health Informatics is about.

The most important aspect is information. Computers and technology simply help us to use to information productively. When you think of Health Informatics, you should think about information, not computers. We will go into the reasons why in more detail in Chapter 2.

So we can slightly revise the ICT acronym to reflect the importance of each component, to

I – Information
C – Communication
T – Technology

This leads us to a definition of Health Informatics:
"The study of information and communication systems in healthcare"
-Enrico Coiera, Guide to Health Informatics, 2nd Ed., Arnold 2003.

Note that technology is left out of this definition entirely!

So now that we know what it is, where do we want to go with Health Informatics?

The 'Holy Grail' is:
• An electronic health record (EHR) – accessible anywhere
• Unique patient identifier - a number that identifies us when we access heath services
• Local, national and international compatibility and access – if you happen to attend hospitals in New York, Dublin and Paris, will they use the same hospital numbers? Will your health information be available when you visit another city? Would you even want this?
• Electronic decision support - will it keep us informed to help us make the correct decisions?
• Online, accessible guidelines and protocols - up to date recommendations

In theory, if we can put all of this in place, it should:
Prevent errors + Save lives
……. Or will it?

What is the evidence for this? How do we improve patient safety?

The famous Institute of Medicine report (US, 2000), "To err is human" noted that there are an estimated 44000-98000 deaths per year due to medical error. This is more than highway accidents, breast cancer and AIDS combined. They didn't even count day case procedures, outpatient cases or nursing home incidents…

In terms of fatalities, it's equivalent to a large passenger aircraft crashing daily. So there is certainly a good reason to improve patient safety.

But, is there evidence to show that ICT can:
-Reduce medical error
-Reduce mortality and morbidity
-Reduce cost of healthcare
?

This book will help you to search for the answers.

The question of whether ICT is of benefit is something that will be at the core of our study of Health Informatics. If there's no benefit, then surely it's not worth all the bother. With millions spent on ICT in healthcare every year, you would expect some benefit, wouldn't you?

Privacy of health information

Have you ever been in hospital? If you have, you might have noticed that all the healthcare team write notes about you in your medical chart. But have you ever read this chart?

Would you want to, if you had the choice?

Even if you didn't, you definitely wouldn't want anyone else to read through them, right?
Most people think their health records are kept in a secure place in hospitals, where nobody can read through them, but in fact most hospital wards keep charts in a trolley in the middle of the corridor!

Imagine you went into hospital for a very personal procedure - (e.g. an operation on your private bits) - who finds out about this? Do all the nurses on the ward know your diagnosis? Do all the doctors talk about your health information? If you needed a physiotherapist after the surgery, would they know what operation you had done?

Unfortunately the answer is that anyone who works in a hospital that uses paper records can easily find out about someone by simply picking up their chart and reading it. There is almost no protection at all.

Case: Information confidentiality
Sheila Martin is a 3rd year physiotherapy student on a clinical attachment. She has just seen an interesting case; a 25 year old local man who presents with leg weakness. Sheila performs an assessment with the doctor, and they decide to arrange a specialist referral for a possible diagnosis of MS (multiple sclerosis). The patient is not yet aware of the possibility.
Sheila meets her sister and her fellow students for lunch in a cafe where she tells them about the case in detail.

Self-reflection
Do you think this represents a breach of confidentiality, or legitimate discussion? Why?

In this book we will look closely at how information is collected in healthcare, how we should be protecting it, and how electronic records will hopefully help.

Summary
Health informatics is about information and communication in healthcare

Technology is just a tool to help us control and manage information and help us communicate
Privacy of information is very important in healthcare and electronic records should help us to improve this

Chapter 2: Data, information and knowledge

The three concepts above are important. The terms are often used interchangeably, but the meanings are different.

Take this example:
What do these numbers say to you?

38.1, 39.9, 42

Does your mind attach any significance to them immediately?
If there is no meaning to them, the numbers are useless.
They are just data.

But what if we give them some relevance?
e.g. if we say they are a sequence of temperature readings in Celsius on a TPR (temperature, pulse, respiration) chart

Now we have information - which is potentially very useful. This is data in context and with perspective. We can interpret the series of numbers, and say that the person has a fever, and the temperature is rising.

Very well, but don't stop there!
Knowledge is an even higher concept. It is information, with guidance for action.

So this would obviously be a sign to intervene. Maybe do a septic screen, give something to take down the temperature and possibly start treatment...

The hierarchy of data, information and knowledge highlights the fact that we need to use the correct terminology depending on what we are dealing with. Plain data on its own is not often useful. Any of you who have done research may have collected massive amounts of data, then you need to try and make sense of it - i.e. turn it into information. By interpreting it you can increase your knowledge.
(This book aims to increase your knowledge, by the way...)

Case study:
Dr Wings is a surgeon in Main River General Hospital. The new administrator begins a scheme whereby all the surgeons' morbidity and mortality data is posted on the hospital website. The aim is to increase transparency and empower patients.

Unfortunately for Dr Wings he specializes in surgery for older people, and as a result has a much higher mortality rate. His patients are sicker than others, and more complicated procedures are required.

The data is posted in its raw form and this causes damage to Dr Wings' reputation. Patients are afraid to be operated on by him, and the local newspaper expresses concern.

This example shows how data in its raw form can give the wrong impression. It would be important to know the average age and other characteristics of the population he operates on, as well as details of the types of procedure and other factors. Information is required here, not raw data.

Analysing data
So is all data good quality? In fact how do you know what data quality is?
Have a look at these characteristics:

Accuracy
Precision
Comprehensiveness
Currency
Relevancy
…

Really this looks incredibly boring, but there are certainly lots of ways to describe data – even more than on this list. Would you think about all this (or any of this) when you record healthcare data?

Collecting data
What types of data/information do we collect in healthcare?
And how do we collect it?

Here are some examples:

Non Clinical data	Clinical data
Name	Presenting complaint
Age	Medical history
Address	Medications
DOB	Allergies
Gender	Vital signs
Marital status	
Religion	
Sexual preference	
…	

Some of these pieces of data can be quite personal - not necessarily something you want to talk about in public. So data is still something that needs to be well secured (not just information). Do you think we really need to collect this much data?
In fact, have you ever thought of the sheer volume of data that is collected on patients? Many different people in different healthcare sites enter data. It creates huge files, which are not always linked together.

Here is a diagram of how data might be collected on you if you go into hospital...

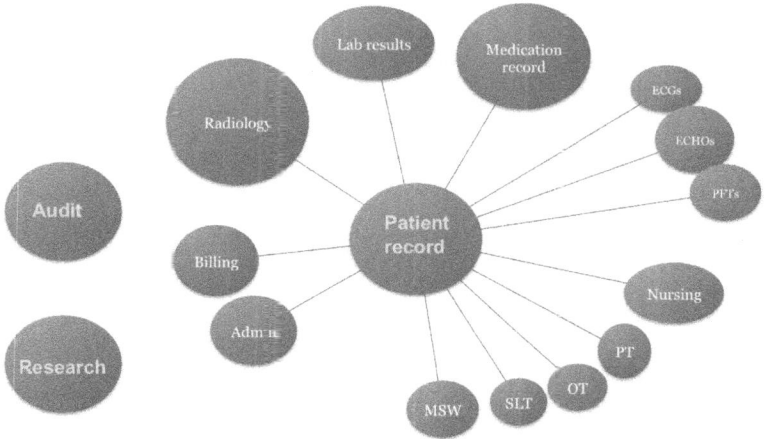

There are possibly hundreds, or thousands of entries made for each patient admission. Of course, several systems may replicate the data. You might be asked to give your name and date of birth again and again on admission - to the nurse, doctor, administrator and radiographer... And of course you might meet several doctors, all asking similar questions, and all writing separate notes of varying quality and usefulness. You may already have noticed this.

We are very good at collecting vast quantities of data, but what if you need to find out something from the pool of data?

The Irish health service provider, the Health Service Executive (HSE) was recently asked to report on how many children had died in State care. This seems like something you should be able to answer with a few clicks of a button. But it took several weeks before preliminary figures were released, which were then adjusted to the final correct figures later. This raised questions about the management of information by the health service in an extremely sensitive area.

This case illustrates how informatics is vital to the running of an efficient, effective, transparent health service.

Self Reflection
1) Have you ever seen data misused?
2) Have you seen confidential data in an inappropriate place?

Data safety and confidentiality

When you think of a health professional entering notes on a patient, how do you imagine it? Sitting at a desktop computer? Using an iPad? Maybe on their phone? Surprisingly many hospitals and practices still use paper records. Imagine you were a health professional, writing notes in someone's chart when you spotted another chart next to it with a name you recognised. It is your father's friend from home. Instinctively you would probably want to know why they were in hospital. And as a healthcare worker nobody would question you for picking a random chart and flicking through it. But how entitled are we to know about other people's health? Let's look at a high profile case.

Case history: Publicising private health data
In December 2009, the TV3 channel in Ireland featured a news item claiming that Irish Minister for Finance Brian Lenihan had been diagnosed with pancreatic cancer. This was despite the fact that he had not made this public knowledge, and had not given consent for his health data to be shared. There was a strong public reaction, and complaints were made to the Broadcasting Authority of Ireland. TV3 justified it by saying it was in the public interest. The complaint was ultimately rejected. The Authority agreed that it was ok for TV3 to broadcast it. Mr Lenihan later made his own statement confirming a diagnosis of cancer.

See the original broadcast here:
http://www.youtube.com/watch?v=ZlrqD6YyEOs)

Irish Independent report on public reaction:
http://www.independent.ie/national-news/public-anger-at-tv3-intrusion-into-lenihans-serious-illness-1990191.html

A similar controversy had erupted in France soon after President Mitterrand's death in 1995. Mitterand's personal physician wrote a book about the president's fight with prostate cancer - something the public had not been aware of. He was charged with breach of professional secrecy and sent to prison!

Self reflection:

Do you think it is in the public interest for us to know about a politician's health, or should it be a criminal offence to make personal health information public?

Interestingly, in the US the President always makes his medical report public! You can see it here: http://blogs.suntimes.com/sweet/potusmedicalexam.pdf

Here is an example of someone who found out their health information was sold:
(it's a long article - just read the first few paragraphs)
http://www.nytimes.com/2009/08/09/business/09privacy.html

One hospital recently found out that confidential patient data had been leaked. They had outsourced the typing of dictated letters to a company in the Phillipines.
http://www.thejournal.ie/tallaght-hospital-admits-patient-medical-records-were-breached-192140-Aug2011/

Can you imagine if you went into hospital to have a sensitive matter dealt with, and it appeared on the internet? How can this happen - aren't there laws to stop this?

In the European Union and the United States there are laws in force to protect personal data. In the US in particular, there are severe penalties for people who breach health data codes. In Europe most law is based on Data Protection, which is not specific to health data. Many people argue that health data deserves even stronger protection because of its sensitive nature. It seems likely that this area will be strengthened with more widespread use of electronic records.

Summary
Data is the lowest level – it is plain facts or figures
Information is data with context
Knowledge is information with guidance for action
Health services collect huge quantities of data
Strong protection is needed to ensure privacy and security

Chapter 3: Information Systems

We are now going to look at bit more closely at how we attempt to organise information in our clinical practice. This will teach us a bit about the building blocks for an EHR.

Information systems

Back to data, information and knowledge

Our example from last time was simple:
Data: 37.5, 38.2, 38.8, 39.1

When you see them in context on a chart, you now have information:

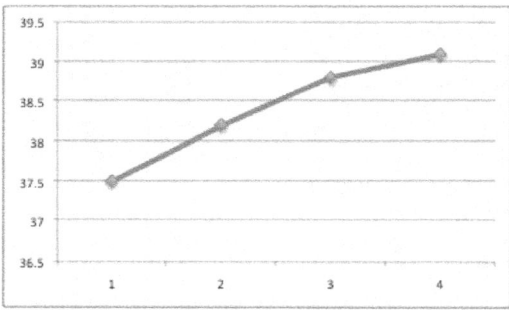

Temperature chart Bill O'Hare 651384

And you can then interpret this information and make a plan of action.

You are familiar with clinical monitors, often seen at the patient's bedside. These track changes in heart rate, blood pressure and other measurements and attempt to interpret the information. They do this by using systems.

What is a system?
A system is anything that takes something in, processes it and gives out a result

An excellent example is the human body:
Input - food
Processing - digestion
Output - poo

There may also be feedback, which allows the system to self-regulate. For example when your belly is full you reduce the amount that is input.

This simple concept of a system is important, because they are the basis for many more complex structures in medicine.

Information Systems are systems which handle and process information. They are designed to manage a set of activities, such as processing a blood sample or deciding staffing levels on a unit.

What is a health information system?
A health information system is composed of numerous information systems and applications (software programs) which are joined together in a coherent way. Some of the following information systems may be components:

Laboratory information system
Pathology information system
Picture archive system
Patient billing system (most of the first hospital information systems were developed to streamline billing)
and so on...

Information systems are not new - they existed before computer technology. There are still many paper-based ones, such as the paper clinical record. Indeed some should not be computerised - a simple system may work better on paper, rather than formalised on a possibly overcomplicated software program.

Databases
The information systems in healthcare need to have access to good quality data. We saw in chapter 2 that there are many different ways of looking at data quality. No matter how well designed the information system in a hospital is, if there are problems with the quality of the data then it will be unlikely to work well.

So to start off, the data must be of a high quality. Now let's look at how databases can help to organise it.

Database structures
In a database the data structures identify both the data being stored and also the way in which the data is linked to each other. For example, in the first ever version of a database, there was a hierarchical structure, with each data item having one or multiple "offspring", but only one "parent" - see figure. This was called a hierarchical database.

hierarchical database:

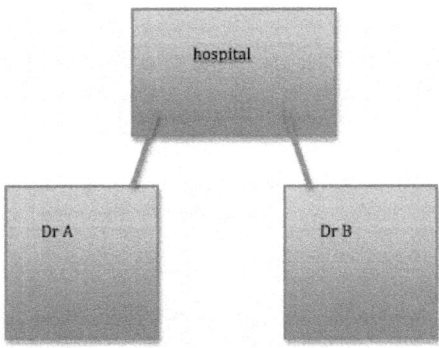

But what if a doctor was attached to more than one hospital? You would need a completely separate database to show this. So the hierarchical database developed into the network database, where several parents were possible. (You don't need to remember these names, just the fact that there was a sequence of development of the types of databases)

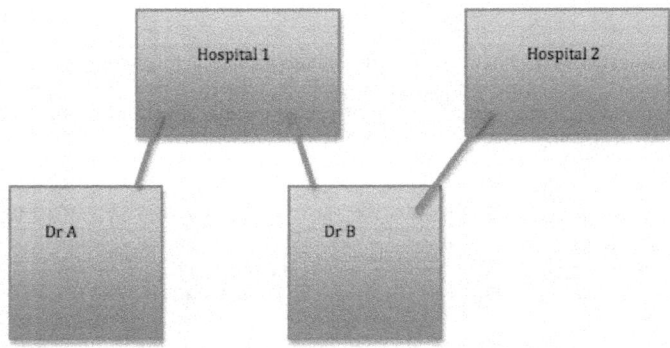

However the coding and maintenance of these databases became very complicated as the database increased in size, so that another way had to be found.

Relational databases simplified the above methods by having a set of two or more tables, linked together by a single common field - one which was shared by both tables, and allowed the user to shift from one to the other.

Table 1

Dr ID number	Dr Name
10511	Dr Nolan
10512	Dr O'Brien

Table 2

Hospital name	Dr ID Number
St Instance	10511
Central	10512

Besides simplicity, these tables also have the advantages of greater security and the ability to restrict access to part of the database, depending on user privileges. For example there may be a Table 3 which lists the salary of each doctor and this information would only be available to the relevant department.

Does this type of database look familiar to you? It is the type used by most common software programs today, such as Microsoft Access. (A program such as this is called a Database Management System or DBMS)

Relational databases represent a huge advance from the early days when each hospital department had their own flat file database - essentially just a list of data with no links to anywhere else. With relational databases each department can link their tables to the others. This can then be pulled off as a single record for each patient, comprising information from each different department. This should mean that as a patient, you are not asked the same question three or four times - when your name and details are entered at the hospital reception, the x-ray department shouldn't need to collect the same data again.

Why do I need to know about databases?
In your research and audit activities as a health professional you will be making databases to organise and analyse the data you collect. It is important that you have a basic understanding of how databases function for this.

The Clinical Data Repository
A database is also known as a data repository. In healthcare, the Clinical Data Repository (CDR) is the name given to the space in which all clinical data is stored. As we saw earlier, data may come from several different hospital departments, such as laboratory, pharmacy and radiology. The CDR is the central area for keeping all this valuable data.

The idea of the CDR is to have one place where all useful data can be held - but imagine what would happen if each department uses a different language or classification system? There will be no interoperability of the systems. In this case, a Clinical Data Dictionary (CDD) is required to translate the messages being sent into one common form, which can then go into the CDR. Equally when messages go in the opposite direction they will need translation.

The above obstacle exists where "legacy systems" form part of the health information system. For example, the laboratory system may be very old and use outdated terms for the data that it stores - this may be entirely different to the terms used in the more modern radiology system. Rather than throwing out both and starting from square one, you can use a CDD to act as an interpreter. This saves money and effort, but might limit future growth and versatility of the system. It can also slow it down.

Many hospitals use this system, called "Best of Breed" - each department bought what it thought was the best system for the money at the time. Each system now forms a component of the whole information system (sometimes called the Hospital Information System). Unfortunately in some hospitals the components still do not communicate with each other so you just have a group of separate systems in each department. It's like the earliest version of the databases we saw - they never got them linked up in one relational database. This often necessitates logging into one system to check lab results and another to check x-ray results (an everyday reality for me - in fact if I want to look at an x-ray I log into one system and if I want to see the report of the x-ray I need to log into another!). A CDD could fix this, but is there a simpler way?

The Unified Database
This is a fully integrated system, i.e. one whose components were built concurrently and can speak to each other without an interpreter (so do not require the use of a CDD). This system is easier to set up, maintain and repair. However it has a much higher cost and means that any legacy systems (and worse, possibly even the data on them) are now defunct. Another limitation is that all products come from the same manufacturer (usually a software company), and if they run into financial or technical problems the project may be put at risk. In the US there were instances where software companies went bankrupt, and hospitals lost not only the ability to use the programs, but also the data being stored. As you can imagine, this could be catastrophic.

Newly built hospitals have the ability to design their health information system from the ground up - so often put a unified database in place from day one. They don't need to include any old data, and they want a simple system with just one number to call when something goes wrong. All their systems talk to each other. In the long term it saves money and yields a return on their initial large investment.

If a patient goes from their GP or family doctor to a private Emergency Department, then is transferred to an Intensive Care Unit in another hospital, any electronic data should follow them. Often, though the data is entered electronically it is then printed out to be transported by hand to the next provider... who then enters it into a computer again.

Back to the Health Information System
Before we branched off to talk about databases we started looking at health information systems. We saw the components, based on which department they were in (lab, x-ray etc). Now we will take a look at how the components should be put

together, but with a functional viewpoint. This is the ideal list of what should make it up:

Components of the Health Information System (a functional view)
Computerised Physician Order Entry (Ordering tests and viewing results)
Decision support
Clinical documentation (including sections for doctors, nurses, physios, OTs, etc)
Picture Archiving and Communication Systems
Electronic Medication Administration Records
Ambulatory Electronic Health Record Systems

Outpatient Systems:
Disease registries and preventive medicine
Two way lab resources
E-Prescribing

Why do we need good information systems?
We saw in Chapter 1 that humans are prone to error, and that medical error accounts for probably 100,000 deaths per year in the US ("To Err is Human" report). The Institute of Medicine launched an effort to rectify this in 2001, with their report "Crossing the Quality Chasm A New Health System for the 21st Century". They set up a Quality Initiative to study these issues:

   * How the experiences of patients should be changed
   * How teams of healthcare workers should interact
   * How healthcare organisations can better work and institute proactive error-reduction strategies
   * How policy officials and healthcare purchasers can reshape health policy to create a safer healthcare system.

You can see the document here:
http://iom.edu/Reports/2001/Crossing-the-Quality-Chasm-A-New-Health-System-for-the-21st-Century.aspx

One of the core statements is that "Patients should have unfettered access to their own medical information and to clinical knowledge. Clinicians and patients should communicate effectively and share information."

Self reflection: Access to your medical records
Do you think you should have access to your own medical records?

(Currently in many countries if you want to view your record, you need to make a Freedom of Information request. This can take several weeks. The exception is in many maternity hospitals. Pregnant women are given their chart to keep at home, so that they can bring it with them if they happen to go into labor when they are near a different hospital.)

Can you see any difficulties with people having free access to their records?...How would clinicians feel about patients reading what they have written about them?...

The IOM also called for design of new systems that "prevent, detect, and minimise hazards and the likelihood of error." They wanted to make a "system in which it is hard to make a mistake and easy to do the right thing."

Sounds like a nice idea. But how can we achieve this?

Improving information flow
At this stage you will appreciate the huge importance of ensuring information flows well in healthcare. The IOM recognised this, and stated that "information technology must play a central role in the redesign of the healthcare system if a substantial improvement in quality and safety is to be achieved".

It is clear that everybody agrees that improving information flow is central to a better health system - but even reading the above statement, you can see that there is a huge focus on the technology rather than the information system itself. Remember an information system doesn't even have to be computerised to be effective. Many projects in the last few years spent huge amounts of money on hardware and software but didn't actually satisfy the central aim of improving the information system. So they look nice but they don't work any better than existing systems... It's easy to get distracted by flashy graphics but the crux of the matter is how well the information gets to where it is needed.

By studying health informatics you can raise your awareness of these issues. Most of you probably want to be able to use EHRs if you don't yet. If you don't speak up and demand that any EHR suits your needs, you will either be using a system that is badly designed for your needs, or still using paper!

Here is an up to date example of an EHR that is multiprofessional - doctors, nurses, physios all use the same record.
http://www.bcmj.org/blog/paperless-ehr-first-team-canada-2012-olympics

Summary
Information Systems are systems which handle and process information.
The aim is to have a unified database that controls all clinical information
Information systems should prevent, detect, and minimise hazards and the likelihood of error
They should be well-designed so that they improve information management

Chapter 4: The Clinical Record

For years doctors have been writing notes to record what happens during a patient visit, and now all health professionals do. This was initially used as a reminder and to keep track of progress. Nowadays they are also important for other reasons, such as medicolegal purposes, audit and research. Their complexity has increased over time - with more sections being added and the size increasing from a simple file to a large folder or chart.

Functions of the Clinical Record:
Provides a means of communicating between staff managing a patient, who may be on different work shifts or may not see each other during the day
Serves as a single access point for all information relating to the patient
Offers an informal working space to record the clinical impression and plan
Serves as an archive or record of care after discharge, and for long term follow up

Self-reflection:

What are the advantages and disadvantages of paper records?

Feedback:
Advantages: Portable, familiar to everyone, can get a quick overview by flicking through pages

Disadvantages: Can only be used by one person at a time, no backup if lost or damaged, take up space, can be heavy and cumbersome, possibly damaging to the environment.

Information entry in the paper record
One of the great advantages of a paper record (for some people) is the lack of structure. You can write your entry any way you like.

You have seen clinical records and will know that a variety of things differ depending on who entered the record:

Language (formal, informal)
Abbreviations
Style

Structure/Layout (or lack of it)
Clarity
Legibility
...

But this variability in entry can cause huge problems - one person may not understand another's entry (this happens quite frequently). It is also difficult to extract data from the paper record. If there is no formal structure, important things may be omitted. If you are looking for a particular piece of information it may be quite hard to find.

How can we improve on this?
With most clinical records being handwritten, is there any point in trying to improve the quality of the notekeeping, or is it better just to wait for electronic records?

In fact, it is possible for a well-designed set of paper records to be superior to a poorly-designed electronic record.

One way to do better is to use the SOAP structure - subjective, objective, assessment, plan.

Subjective - what the patient says
Objective - what the clinician sees
Assessment - your impression of what is going on
Plan - what you are going to do

Another approach is to write "problem lists". This is where a patient's problems are numbered in order, e.g.
1. Multiple sclerosis - declining mobility secondary to muscle spasticity
2. Depression - currently inactive
3. Parkinson's disease - well controlled on medications
etc

Do these structures make notes better or are they an unfair restriction on your freedom to write as you please?

Standards in record-keeping
The above systems aim to improve the standards of records. But, why are high standards important?

A high standard of note keeping is of the utmost importance in clinical practice. Poor notes are found to have made a significant contribution to a very large number of complaints, high profile problems in hospitals and in many cases of litigation which could not be defended. Inquiries in two very high profile scandals in English hospitals in 2008 and 2009 found unacceptably high mortality. Poor documentation at admission and handover was found to be one of the significant underlying factors in both cases.

In 2006, a United States Joint Commission on Health Care report estimated that around 80% of malpractice claims are lost due to inadequacies in the medical record.

25 out of 28 case studies reported by a UK medical defence organisation in 2009 related to record keeping.
20 were settled for substantial sums because of poor record keeping.
5 had good defensible records and the doctor was absolved.

The 2008/2009 Fitness to Practice Report of the Nursing and Midwifery Council reported that errors or omissions in records keeping was the third most frequent underlying factor in inquiries into fitness to practice in 2006/2007.

Among the many benefits to the doctor or other professional, key aspects of good clinical record keeping are:
Improved communication between all the health professionals involved with a patient.
Time saved in finding information – that could be extensive if a patient has had complex problems or has been in hospital for a long time.
Improved diagnosis and treatment because a patient's problems have been clearly recorded and investigated.
All events, decisions, actions taken and relevant discussions are properly recorded and not just briefly and incompletely summarised (or omitted!).
Protection against any subsequent complaints from patients or their representatives.
The creation of an opportunity to learn on the job by being able to properly track how a patient is progressing.

The move to Electronic Health Records (EHRs)
In order to facilitate a smooth transition from paper records to electronic patient records it is useful to introduce standardisation of paper records. Standards for the content and structure of the medical notes are being developed by the medical profession on the basis of consensus.

Medical record keeping standards should ensure that electronic patient record systems will reflect our day to day clinical practice rather than primarily the requirements of IT. If they are organised to meet IT requirements, they are unlikely to be accepted by clinicians as they may be perceived as a hindrance and would be more likely to fail.

In addition to transferring clinically relevant information from the computer system onto the patient records, it is vital to document whether this has been acted upon. All printed reports should be filed in the patient notes.

Task:

The Royal College of Physicians in the UK has done some good work in this area. They are a professional body that helps keep doctors well-trained. Read this guideline that they published:

A Clinicians Guide to Record Standards - Part 1: Why standardise the structure and content of medical records?

Note: This is a 16 page document! You don't have to read everything. While you are going through it, look for answers to the following questions: (read on for feedback, but not until after you answer the questions!)

1) What is the merit to introducing a structured way of entering information into the patient chart now, even before we have an electronic record?
2) Is there actual evidence for a benefit?
3) What are the components in the process of developing the standards that are used?

Task Feedback:
1) Having a well laid out record aids access to the right information. It is much easier to find what you are looking for if there are standardised headings or documents. One of the biggest problems you might have when you start work as a doctor is getting used to how things are done differently in different hospitals. By standardising as much as possible this will make your task easier.

2) Yes, structured records have beneficial effects on doctor performance and patient outcomes. (Page 4 of the document, reference: Mann R, Williams J. 2003. Standards in medical record keeping. Clinical Medicine; 3:329-32.)

3) Review of evidence (Literature review),
   Drafting
   Extensive consultation
   Redrafting (Revision) then agreeing and signing
   Piloting (Deliver)

We will look more at developing standards and guidelines in upcoming session - and it's worth knowing that in your professional life you will be expected to compose guidelines on your area of expertise. So it's an essential skill.

Record Standards

Next read through the standards themselves:

A Clinicians Guide to Record Standards - Part 2 : Standards for the structure and content of medical records and communications when patients are admitted to hospital

Read page 4 in full, then glance quickly through the rest of the booklet. When you have finished complete the Task below.

Task:

Have a look at the clinical notes displayed below and make a list of which standards are not being adhered to, and why (looking at Page 4 of the RCP document)

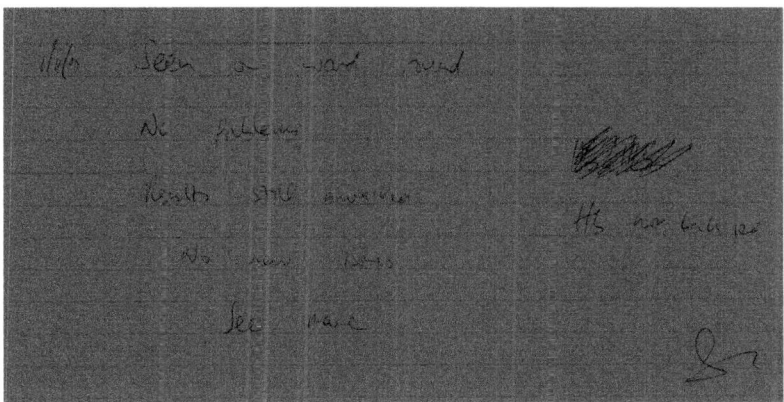

Task feedback:

Standard 2 - no patient name, ID number or location
Standard 3 - no standardised structure
Standard 6 - no time, date unclear, difficult to read, signature not legible. Alteration scribbled out rather than struck out with single line.
Standard 8 - who was the most senior person seeing the patient

Hopefully this illustrates how standards such as those in the document might be able to improve our notekeeping (while we wait for electronic health records)

Summary
These are the main learning points on the importance of good clinical record keeping.

The patient enjoys an improved experience of treatment.
The quality of treatment and safety of the patient are enhanced.
Duplication is avoided and all relevant information is recorded.
Communication between staff is improved.
There should be fewer complaints.
There will be more learning opportunities.

Chapter 5: The Electronic Health Record

The concept of an EHR (Electronic Health Record)
Many terms have been used to describe the same idea - electronic patient record, computer based medical record, electronic medical record...

The current term is electronic health record, and was suggested in the IOM (Institute of Medicine, USA) report of 2003, "Key Capabilities of EHR Systems". It said an EHR should have:

1. Longitudinal collection of electronic health information
(imagine if your childhood vaccinations were instantly on record even though you are now 48 - no need to ask your mother if you ever had chicken pox)
2. Immediate electronic access to person and population level information by authorised users
(So if you go on holidays but get sick, the doctors in that GP surgery or hospital could get your records)
3. Provision of knowledge and decision-support that enhances the quality, safety and efficiency of patient care
(This idea means that EHRs are not just to hold patient information, but also to help clinicians, by teaching them and aiding decisions!)
4. Support for efficient processes for health care delivery
(How can you make healthcare more efficient? By digitising it)

The report also identified eight core areas for which EHR systems should provide supporting features or functions:

* Health information and data
* Patient support
* Results management
* Electronic communication and connectivity
* Decision support management
* Reporting and population health
* Order entry/management
* Administrative processes

Clearly, an EHR is not just an electronic equivalent of the paper record - it should be much more powerful. According to the above list, it should help you to make decisions, manage billing/appointments, handle communications, give patients access to relevant information, and be used in audit and research.

As you saw earlier, each department in the hospital may have their own information system, but the EHR requires that all of these are linked together, usually using a relational database. This would allow you, the doctor, to pull up information on a single screen from several places. For example, ideally you could look at the patient's past history [from clinical record], check if they are on the right medications [pharmacy], find results of their serum drug level [laboratory] and then order a new drug dosage if needed [computerised physician order entry].

This is the core task for electronic records - integration of different information systems.

Data Integration
In order to integrate all components of a system, they will need to use the same consistent scheme for coding data elements, and have a way of transmitting this data around the system. This is usually achieved by having a clinical data repository (which we learned about earlier). Let's have a look at coding in some more detail as this will show you the difficulty of the task ahead for those designing an EHR. The purpose of this section is not for you to know about coding in detail, but to see the massive complexity of getting a system like this up and running. It's a bit technical so hang on!

Coding
Have you ever seen a discharge form that is put in a patient's chart after they leave hospital? It lists main diagnoses, procedures performed, medications, GP details etc. This document is then passed to the coding department to have the diagnoses converted to codes which can be entered into a computer system. For example, in Ireland this is the HIPE system (Hospital In-Patient Enquiry).

You can read about HIPE here:
HIPE
http://www.esri.ie/health_information/hipe/

Also read the page on "Clinical Coding" - click on it in the menu bar on the left of that screen.

You can see from their description of the process that they use the "ICD-10" coding system. There are other systems, such as DSM - in fact there are many in operation.

Briefly read this abstract of a research project:
http://www.ncbi.nlm.nih.gov/pubmed/18005561

In this research, the authors found that for a given condition (e.g. depression) up to 19 codes were shown on screen for a doctor to pick which condition they were referring to. To use the system, every time a doctor enters a diagnosis, they have to pick which code corresponds to it...It's like when you put "John" into your phone's contacts search and find 20 coming up. It really slows you down, and doctors are already pressed for time.

Let's look a bit closer at two coding systems: ICD-10 and SNOMED CT.

Task:
Read about the ICD-10 coding system here:
WHO
http://www.who.int/classifications/icd/en/

From this page, click on ICD-10 ONLINE -> Current version

Look at the list of codes. See if you can find the code for Pulmonary TB (Pulmonary = lung; and TB= tuberculosis), then read the feedback below:

Task Feedback:
You might start by looking at "Diseases of the respiratory system" (respiratory=lung), but actually it's under "Certain infectious and parasitic diseases". (A15.0 Tuberculosis of lung, confirmed by sputum microscopy (this means looking at phlegm under a microscope to see what bugs are in it) with or without culture)

This shows the problem with this type of scheme - a condition can only be classified under one heading, not many. It takes some knowledge of the way the scheme is organised to find the exact term you want. You could actually spend hours looking for the correct code that corresponds to a medical condition.

These coding schemes are called Enumerative schemes - They have one code for every possible medical condition. This is both rigid and has the possibility of duplication as the scheme gets bigger. Imagine - you need a unique code for every possible diagnosis there is... That is at least 6,000 conditions!

Subsequently, Compositional schemes were created because enumerative ones seemed to be getting very large and unwieldy, losing ease of use on the way. A compositional scheme uses combinations of primitive terms to create a more complex term. Essentially they make a dictionary and a grammar, rather than just a phrasebook.

For example the SNOMED CT scheme uses a combination of Concepts, Descriptions and Relationships to create a term. By combining different values for these three entities, a huge number of possible terms can be created.

ICD-10 and SNOMED CT are also called Clinical Terminologies. ICD-10 is essentially a list of all possible diagnoses, so that a computer can "understand" what we mean when we write down a diagnosis. SNOMED CT is much broader than that, it covers many concepts other than just Diagnosis. Have a quick look at the website:

SNOMED CT
http://www.ihtsdo.org/snomed-ct/

Why do we need clinical terminologies?
Capturing clinical data
Information integration, indexing, retrieval
Messaging between software systems
Reporting back on data
It's a hugely complex area and many researchers have already devoted years to creating the perfect version - but no single scheme is recognised as the best.

Health Informatics Standards

Coding is just one of the standards that informatics researchers are trying to develop. Standardising the language that computers use, the terms, the messaging, perhaps even the software and equipment could be very beneficial. It could reduce waste and ensure that we are all trying to do the same thing. But of course there are many different groups proposing standards, (just like there are standards bodies in many different countries e.g. ISO, CEN, NSAI) so which one do we adapt? We will have a look at one here:

http://www.hl7.org/about/index.cfm?ref=common

Read briefly about HL7 on that page, and also the FAQ page.

It's interesting to note that HL7 was mainly developed in the US, and the UK used a different system.

Open Source efforts
One of the reasons you have many different standards is money. There is a lot of profit to be made here - if your standard is embraced that means big bucks. "Open source" essentially means collaborative efforts who allow people to freely distribute their work. In the area of standards, many of the people who write standards charge users for them. But now there is a large open source standards body, OpenEHR. This might allow for a greater level of adoption.

Read about their efforts here: (you can skip the technical bits)

OpenEHR
http://www.openehr.org/shared-resources/getting_started/openehr_primer.html

Usability
Do you think an EHR will be easier to use than a paper record? Unfortunately some are very complex systems and are not very user-friendly. We will look more at how humans and computers interact later. One thing that is sure is that doctors and other clinicians will not embrace an EHR if it doesn't make their life any easier. One of the commonest reasons for failure of any computer system is if it slows down your workrate, or has an overcomplicated interface. Usability studies are where the end-user (usually the clinician) is observed testing the software, and then the software is revised (possibly several times, in an iterative process) so that it becomes good enough for the required tasks.

If you would like to see how EHRs are done in another country, you can look at the website for the EHR in Quebec, Canada (the site is designed to be read by patients, do you think it does a good job?)
http://www.dossierdesante.gouv.qc.ca/en_citoyens_DSQ_et_vous.phtml

Is there any risk to using Electronic Records?

It seems logical that EHRs will obviously improve things hugely. But like everything in healthcare, we do research to see if any intervention actually has a positive effect when it is introduced.

In 2004, an informatics researcher called Joan Ash published an analysis of IT errors that affect patient safety. She warned that errors can occur from the way information is presented, the way IT affects how health professionals interact with each other, and from information overload. (From: "Some Unintended Consequences of Information Technology in Health Care: The Nature of Patient Care Information System-related Errors." Joan S. Ash, Marc Berg, and Enrico Coiera. J Am Med Inform Assoc. 2004 Mar-Apr; 11(2): 104-112.)

The very next year, a very scary study was published in a major journal:

Unexpected increased mortality after implementation of a commercially sold computerized physician order entry system.
Han YY, Carcillo JA, Venkataraman ST, Clark RS, Watson RS, Nguyen TC, Bayir H, Orr RA.
Pediatrics. 2005 Dec;116(6):1506-12.

In this study, the Department of Critical Care Medicine in the Children's Hospital of Pittsburgh looked at mortality before and after they put in place a CPOE system (Computerised Provider Order Entry, in other words, electronic prescribing) to reduce medical errors and mortality. They found that deaths significantly increased from 2.80% to 6.57% of hospital inpatients after implementation of an electronic prescribing program.

This is quite remarkable and it's difficult to say what exactly caused such a dramatic effect. But it's clear that: Information Technology + Humans = sociotechnical change (i.e. changes in behaviour and interaction).

For example, the authors note that because pharmacy could not process medication orders until they had been activated online, ICU nurses also spent significant amounts of time sitting at a computer screen and away from the bedside. This is clearly an unintended effect of computerisation - keeping the nurse away from the patient! Part of the problem of instituting eHealth interventions is that they change workflow patterns that may have been well established for 50 years in a paper-based environment. The consequences are unpredictable.

This site is also worth a quick look for some scary stories:
"Bad health informatics can kill"
http://iig.umit.at/efmi/badinformatics.htm
So you see, there is actually a chance that introducing an EHR can cause people to die...

Summary
EHRs should facilitate longitudinal collection of electronic health information, immediate electronic access to person and population level information by authorised users and provision of knowledge and decision-support that enhances the quality, safety and efficiency of patient care.
Clinical terminologies are necessary for us to accurately digitize health information.
EHRs can cause socio-technical change that is difficult to predict

Chapter 6: Searching for information

Healthcare students and practitioners commonly generate questions about clinical situations they encounter. But how do they do this?

Some interesting research has been done on general practitioners' search habits. One observational study showed that they generated 0.18 questions per consultation. The mean length of consultation was 7.8 minutes, so that's 1.3 questions per hour. They only searched for answers to one-fifth of questions.
González-González AI, Dawes M, Sánchez-Mateos J, Riesgo-Fuertes R, Escortell-Mayor E, Sanz-Cuesta T, Hernández-Fernández T.Information needs and information-seeking behavior of primary care physicians. Ann Fam Med. 2007 Jul-Aug;5(4):345-52.

Another one in 2006 showed that they first consult colleagues, and paper sources.
Coumou HC, Meijman FJ. How do primary care physicians seek answers to clinical questions? A literature review. J Med Libr Assoc. 2006 Jan;94(1):55-60.

A previous observational study found that GPs spent only 2 minutes looking for an answer (but this was in 1999)
Ely JW, Osheroff JA, Ebell MH, Bergus GR, Levy BT, Chambliss ML, Evans ER. Analysis of questions asked by family doctors regarding patient care. BMJ. 1999 Aug 7;319(7206):358-61.

Physicians could not find answers to 41% of the questions they generated.
Ely JW, Osheroff JA, Maviglia SM, Rosenbaum ME. Patient-care questions that physicians are unable to answer. J Am Med Inform Assoc. 2007 Jul-Aug;14(4):407-14. Epub 2007 Apr 25.

Barriers highlighted include:
-the excessive time required to find information;
-difficulty modifying the original question, which was often vague and open to interpretation;
-difficulty selecting an optimal strategy to search for information
Ely JW, Osheroff JA, Ebell MH, Chambliss ML, Vinson DC, Stevermer JJ, Pifer EA. Obstacles to answering doctors' questions about patient care with evidence: qualitative study. BMJ. 2002 Mar 23;324(7339):710.

How do students compare? How long do they spend searching for information, and how successful are they?

In 2010 a survey was done in a graduate-entry medical school on students' search habits. Here are some of the findings:
- The internet and textbooks were used by 97% of students on a daily basis, but only 10% use lecture notes daily
- Students rated textbooks best for quality of information, but the internet best for ease of finding information
- The most popular website was eMedicine; then Wikipedia then UpToDate
- Only 8% use PubMed on a regular basis

- 55% of students use web videos on a regular basis
- Only 2% always cited the source of their information when reporting back in PBL
- Most students spent 21-40% of their study time searching for information
- Half of students had not been trained in how to search for information

(Chaney S, O'Hanlon S, Shannon B, Finucane P. Information Seeking Behaviour in a PBL Curriculum. Presented at the Irish Network Of Medical Educators Annual Conference, May 2011)

How to search for information in 4 easy steps

1   Identify a question
2   Find the best source
3   Search using tools
4   Evaluate the results

Let's go through this step by step, then you can use your own examples.

1. Identify a question

Imagine you generate a clinical question. It might be: How do you treat Multiple Sclerosis (MS)?

Now, make it as specific as you can: e.g., In [adult patients with secondary progressive multiple sclerosis] is [mitoxantrone better than interferon] at [reducing relapse rates]?

Clearly that question would be made by someone who knew a lot of specifics about MS! It's just to illustrate how detailed the question should be.

When searching some databases you can use specific terms to help your search

2. Find the best source

Medical sources – some examples:
Books
Journals & Review articles
MEDLINE, EMBASE
UpToDate
Cochrane collaboration
eMedicine
NICE
Government publications

3. Search using tools

For example, there is a great database of several MILLION biomedical journal articles, called MEDLINE. To search the MEDLINE database you can use different interface options:
PubMed
OVID
Medscape
Infotrieve
PaperChase

Unfortunately, there is a common problem:

Information Overload
PubMed is the website that most health professionals use to find journal articles. If you type random terms into PubMed you often get a massive amount of information. Clearly most of it is useless. You have to be more specific, and find the exact information you are looking for.

4. Evaluate the results

Is the information current?

Is the source reputable?

Is the question answered?

Successful searching
What we want is a manageable amount of output, and hopefully the answer you are looking for is in there.

Successful searching is:
A balance between recall and precision

More recall=
    more irrelevant results

More precision=
    less results, possibility of
    missing some relevant ones

Searching PubMed
It's very useful to learn how to use the PubMed website to search the biomedical literature.

http://www.pubmed.com

Here's an example:

Find any randomised controlled trials (RCTs) relating to MSA (multi system atrophy) in the elderly

There are many possible ways you could search for this, e.g.:
MSA elderly
But this currently produces 1482 results
(seems like too many to read through)

Multi system atrophy elderly
= 53 results
(great - a smaller number, but it includes other types of "atrophy")

"Multi system atrophy" elderly
= 12 results
(surely the jackpot? No - none are RCTs)

MeSH terms
This stands for Medical SubHeading.
A team of researchers codes every scientific publication, and assigns codes to it to describe the content. Every major concept in medicine has a MeSH term.

On the Pubmed homepage, click on "MeSH database" for more information, including a tutorial on MeSH

When you search using a MeSH term, you can be sure that the articles you find will indeed have been tagged with that subject.

One of the other very convenient features is the ability for Pubmed to monitor any new articles that would turn up in your search query as time goes by. You can do this by opening an NCBI account and going into Saved Searches. When you save a search you can then ask Pubmed to keep you notified with any updates. If you are doing a literature review, you can click on interesting abstracts that appear in your search results, and "Save to Collection" so you can only keep the ones you are interested in.

There are excellent video tutorials on all aspects of searching on the PubMed website. See the PubMed tutorials link on the homepage. If you work out how to use it, it will save you days or weeks, not hours....honestly.

Just remember that your search needs to be focused enough to limit the number of results, but broad enough not to leave out important ones.

Why do I need to know how to use Pubmed? I prefer eMedicine
If you are looking for the answer to a clinical question, you can certainly check eMedicine or another website that aggregates information for you. However these are secondary sources - Pubmed lets you find the original articles yourself and ensures that you don't miss anything. It also allows you to perform targeted searches which may be more specific than what you will find on other websites. Finally, when you do research you will need to perform a review of the literature on your chosen topic.

So it is reasonable to use other websites on a regular basis, or for a quick look-up. But for a detailed search, or for your own research, you will have to do the manual search through Pubmed yourself. If you use the method above you will save a lot of time!

Task: Your own PubMed search

Go to the PubMed site, and search for the answer to a clinical question.

Make notes as to what search terms you use, and how many results you get. Were you successful?

Barriers to Searching, and some solutions
Inadequate search strategy - Get the strategy right before you start, use MeSH
Restricted access articles - Some journals are Open Access (free to read)
Out of date or irrelevant information - Use Limits
Poor quality information - know how to assess information quality
Information quality

This is an example list of indicators of information quality. A study in 2002 found that there were 86 different quality criteria in practical use.
Eysenbach JAMA. 2002;287:2691-2700

It's clearly very subjective and so is difficult to assess. We were wondering earlier whether Wikipedia is an appropriate source - how does it do according to the above criteria? Unfortunately the authority, validity and objectivity are not always clear. It might be brilliant, but there's a risk it's complete rubbish. Controversial, perhaps?

Most clinicians are not trained in how to assess information quality, yet we need to do this all the time. If you see a journal article, how do you know if you can apply the findings? Can you trust the authors? Does the place of publication reassure you? Unfortunately it can take time before people realise that their judgement was wrong:

Have a quick read of this article in the Guardian about the Wakefield paper on MMR published in the Lancet, and allowed to stand for 10 years.
http://www.guardian.co.uk/society/2010/feb/02/lancet-retracts-mmr-paper

If clinicians are not necessarily good at deciding what information to trust, how can our patients be?

Consumer informatics
This field is a subspecialisation of Health Informatics concerned with how health service users access information.

Have you ever seen someone bring a printout from the internet in to their doctor? Maybe you have done it yourself.

Or maybe you have done an internet search on a disease you thought you had (don't worry, it's a common medical student preoccupation).

If you type "Cancer" into Google you get 180 million results. What use is that? Do we need to train our patients on how to search for information, and assess its quality? Or should we just point them in the right direction?

The flow of information on the internet is massive, and sources that are not reputable frequently come up high on the list.

Read this abstract from Archives of Disease in Childhood:
http://adc.bmj.com/content/95/8/580.full

The authors searched for answers to five common paediatric questions, and found that less than 40% of sites gave correct information. The accuracy of information varied hugely.
They found that governmental sites were very trustworthy, news sites only half the time, and no sponsored sites were.

How do consumers search?
Eysenbach and Kohler observed people searching for certain terms. They found that:

-Consumers successfully find an answer to their question, in an average time of 5 mins 42 sec. (they didn't check if it was the correct answer)
-They primarily looked for the source, a professional design, a scientific or official touch, language, and ease of use.
-No participants checked any "about us" sections of websites, disclaimers, or disclosure statements.
-Very few remembered which websites they had retrieved information from.

Eysenbach G, Kohler C. How do consumers search for and appraise health information on the world wide web? Qualitative study using focus groups, usability tests, and in-depth interviews. BMJ. 2002 Mar 9;324(7337):573-7.

There are many sources of information for patients - medication inserts, health education leaflets, the internet, and the media. The same problems of information quality apply everywhere.

Efforts to ensure quality
HON (Health On Net) – a voluntary code of ethics
Ensuring Quality Information for Patients (EQIP) tool
DISCERN tool - evaluation of evidence-based content of health-related websites
Guidelines for patient information
Readability scales (SMOG, Flesch, etc) - to ensure the language level is appropriate

At the end of the day, it's impossible to regulate the internet when it comes to health information. If your patient asks for more information, the best thing to do is to recommend an official government site, or another trustworthy site that satisfies as many of the quality criteria above as possible.

Summary:
Many clinicians find it difficult to search for information to answer clinical questions.
This chapter has introduced you to the search engines that health professionals use when they need to find high-quality information. This gives you an idea of how to do very focussed searches, looking for original scientific articles.
We also need to be able to guide our patients to find relevant high-quality health information

Chapter 7: Online professionalism

Professionalism

All clinicians are expected to behave professionally, and professional bodies provide guidelines for good behaviour. Patients expect us to be professional. Healthcare students' professionalism also has an impact on their future ability to practice. One study has shown that unprofessional behaviour in medical school is associated with future disciplinary action as a doctor, and was a stronger predictor of this than having low grades.

Papadakis MA, Teherani A, Banach MA, et al. Disciplinary action by medical boards and prior behaviour in medical school. N Engl J Med. 2005;353(25): 2673-2682.

Part of professionalism is knowing how to maintain professional boundaries in day to day clinical practice. For example, becoming too close to patients can obviously have a detrimental effect on the doctor-patient relationship, and can even get you struck off.
http://www.dailymail.co.uk/news/article-1308690/Harley-Street-psychiatrist-Theodore-Soutzos-struck-off.html

For many people, going online is clearly personal time, and a break from the day to day grind of study and work. It seems unlikely that what we do or say here would ever be examined. But of course many people have met patients online, or have received friend requests from them. This can be a tricky situation.

Interacting with patients online

Many doctors now use email and online messaging to keep their patients up to date. With a online personal health record, it's possible for patients to receive their lab results, or arrange appointments online, and request a repeat prescription. In most cases there is no ethical dilemma with interacting with patients in this way. But what about potential interactions on social networking sites such as Facebook or Twitter?

It seems likely that these interactions would be frowned upon by professional bodies. With the potential to have serious negative consequences for the doctor-patient relationship, ethicists now advocate avoiding any contact with patients in this way.
Guseh JS 2nd, Brendel RW, Brendel DH. Medical professionalism in the age of online social networking. J Med Ethics. 2009 Sep;35(9):584-6.

Who reads your posts?

Another difficulty is that what you post on Facebook or other sites may be more public than you realise. Many people wrongly assume that only their friends can view their profile or posts, but they have not set up vital security settings to ensure this is the case. Unfortunately there are well-publicised examples where doctors revealed patient information online.
http://www.news.com.au/technology/doctors-caught-revealing-secret-information-on-facebook/story-e6frfrnr-1225929424789

Case: Social Networking

Jenny Moroney is a third year medical student on placement with a family doctor in a rural area. She has settled in well and has developed an excellent working relationship with several patients who regularly attend the practice.

One morning she meets Mrs Pitt, a 51 year old lady with asthma. Having completed her consultation she invites the doctor into the room, and Mrs Pitt's daughter Becky also asks to enter. Becky immediately recognises Jenny's face "You look really familiar, have we met before?" Jenny looks blank, then suddenly Becky starts giggling and says "You're that girl with the Facebook group on 'How to cheat in your medical exams – you are legendary!"

Becky continues: "I've applied to medical school for next year and I was googling options when your group popped up. That's so cool that you can download all the exam answers from your page. Should make life easy for me!"

Jenny looks uncomfortably at the doctor.

Privacy settings
Could your online activity get you in trouble with your medical school? A 2009 survey of US medical school deans assessed their experience with online posting of unprofessional content by students.
Chretien KC, Greysen SR, Chretien JP, Kind T. Online posting of unprofessional content by medical students. JAMA. 2009 Sep 23;302(12):1309-15.

Sixty percent of US medical schools responded (78/130). Of these schools, 60% (47/78) reported incidents of students posting unprofessional online content. Some examples included:
Violations of patient confidentiality were reported by 13% (6/46).
Student use of profanity (52%; 22/42)
Frankly discriminatory language (48%; 19/40)
Depiction of intoxication (39%; 17/44)
Sexually suggestive material (38%; 16/42)

Of 45 schools that reported an incident and replied about disciplinary procedures, 30 gave an informal warning (67%) and 3 reported student dismissal (7%)

The impact of this is huge - if as a student you have a publicly viewable site with any of the above types of material, there seems to be a potential for being disciplined. Is this fair? What happens if someone else posts a picture of you looking a little merry? What if you're not even on Facebook, but your pictures are? Can you even delete pictures from Facebook?

At one institution, teaching about how to elect privacy settings on Facebook resulted in an 80% decrease in publicly accessible accounts.

Coutts J, Dawson K, Boyer J, Ferdig R. Will you be my friend? Prospective teachers' use of Facebook and implications for teacher education. Paper presented at: Proceedings of Society for Information Technology and Teacher Education International Conference; 2007; Chesapeake, VA.

Medical Students' Use of Social Networking

One study examining medical students' use of Facebook revealed that nearly two-thirds had a personal profile and regularly use Facebook. Only 38% made their Facebook sites private. 80% had a personal photograph available publically, and 88% an email address. Sexual orientation, relationship status and political views were freely viewable in over half.

Thompson LA, Dawson K, Ferdig R, et al. The intersection of online social networking with medical professionalism. J Gen Intern Med 2008;23:954-7.

In a more recent study in New Zealand a quarter of recent medical graduates had unsecured Facebook pages. The most frequent unprofessional content was related to alcohol and offensive language.

MacDonald J, Sohn S, Ellis P. Privacy, professionalism and Facebook: a dilemma for young doctors. Med Educ 2010;44:805-13.

Over half of Photo pages showed the doctor using alcohol and 10% of photos suggested intoxication. Some pages had photos of patients; subjects making obscene gestures, cross-dressing or showing nudity. In 22% of cases there was membership of groups that may potentially discredit the medical profession, e g. 'perverts united'.

In focus groups, medical students have spoken about how they maintain privacy on Facebook:

'I like keeping it completely separate ... I have a lot of people I know in the course; I don't have very many medical friends on Facebook because I want to keep it completely separate and people can't find me because ... I know that can affect your career later, so part of me wants to quit [Facebook] anyway.'

Finn G, Garner J, Sawdon M. 'You're judged all the time!' Students' views on professionalism: a multicentre study. Med Educ. 2010 Aug;44(8):814-25.

Your digital footprint

As you can see, your personal (and potentially professional) reputation extends online and may even be beyond your control. What type of personal information (regarding yourself) is available online? This is your digital footprint.

A 2010 study searched for 250 randomly selected physicians registered in Massachusetts.

Mostaghimi A, Crotty BH, Landon BE. The availability and nature of physician information on the Internet. J Gen Intern Med 2010;25 (11):1152-6.

92.8% had professional information and 32.4% had personal information available online. Among personal sites, the most common categories included social networking sites such as Facebook (10.8% of physicians), hobbies (10.0%), charitable or political donations (9.6%) and family information (8.8%). Physician rating sites were identified for 86.4% of providers, but only three physicians had more than five reviews on any given rating site. The authors concluded that physician information is widely available on the Internet, and *often not under direct control* of the individual physician. They advise that physicians should monitor their online information.

Of course patients have online personae too - is this something we should examine?

Case: On Examination, large footprint

Sydney Oz is a doctor training in Emergency Medicine. He is asked to see Bleak, a 16 year old boy brought in after his friends notice him becoming extremely anxious and paranoid. After initial review he suspects drug abuse but the boy denies using any illicit drugs. While awaiting the lab results, Sydney googles the boy's name. On Bleak's twitter account he sees an entry about "going for some kicks tonight with the guys". He then looks at his Facebook page and sees several pictures of him posing while smoking and drinking hard liquor.

Sydney tells Bleak's parents about his investigations.

Indelible
As you can see, the web is like a continuous conversation - but unlike talking, anything you contribute online may be recorded for posterity. Did you know that http://www.archive.org maintains old versions of websites? What if you wrote something silly on a website back in 2003 and it's still up there? Scary, isn't it?

Newer Web 2.0 applications such as Youtube allow sharing of content such that things are now available online that never used to be. Medical student shows (such as a Christmas panto poking fun at staff), can obviously be very funny - but if made publicly available, can cause significant distress and public impact.

Solutions, quick!
Until very recently, there were no guidelines on how to ensure online medical professionalism. So if you are feeling guilty, don't worry. Many people are not aware of the problems highlighted here. This clearly needs to change. Luckily there is emerging guidance on this.

The Australian and New Zealand Medical Associations (AMA and NZMA) and Medical Students' Associations (AMSA and NZMSA) have published "Social Media and the Medical Profession: A guide to online professionalism for medical practitioners and medical students". Several other professional bodies and groups are drawing up their own guidance.

Also, this is a recently published guideline for doctors regarding their online behaviour. It also neatly sums up all of the points discussed in this chapter.
Landman MP, Shelton J, Kauffmann RM, Dattilo JB. Guidelines for maintaining a professional compass in the era of social networking. J Surg Educ. 2010 Nov-Dec;67(6):381-6. Epub 2010 Nov 5.

1. Monitor your personal reputation
2. Understand the privacy settings of the websites you use
3. Remember your audience (intended and unintended)
4. Be aware of the permanence of online content
5. Maintain professional boundaries

Summary
Social media offers many new possibilities for health professionals and students. However the possibilities for adversely affecting their reputation are significant. Education on these issues is vital for healthcare students and professionals so that they may use the benefits of social media in a way that does not impact on their professionalism.

Chapter 8: Communication

It may seem a bit odd learning about communication by reading. We'll be looking a little bit at the theory behind it.

Back in Chapter 1 we used the acronym ICT to describe health informatics: Information, Communication and Technology. We've looked at information in detail, and will see some examples of useful technology later. So where exactly does Communication come into it?

Transmission of information

Most people would say that communication is something to do with getting your message across - i.e. transmitting information from A to B. That is certainly the basic idea, but then if it's so easy why are people sometimes so bad at it?

Let's examine some detail behind the process of communication and see how it can go wrong.

Message transmission

The above diagram shows a basic representation of communication. A Sender sends a message, through a Channel, and a message is received by the Receiver.

Notice however that the messages are named differently - Message 1 and Message 2. This refers to the fact that the message you send is never really the exact message that is received.

Why? It's like Chinese whispers - there is some "distortion" between the message being sent and received. This is one of the most unnoticed principles of communication - everyday we tell people things and assume they get the message perfectly, just as we mean it. Unfortunately it rarely happens that way.

Imagine a hospital ward - it's busy: there are lots of patients, doctors, nurses, medical students, nursing students, porters, ward clerks, and you. You are trying to assess the mobility of a patient that came in last night. As you speak to them, a porter is maneouvring a trolley around you, a team of 7 doctors are discussing an x-ray, the catering staff are delivering lunch, a cleaner is vacuuming the floor and there are 2 families waiting to talk to the nurses.

Now in the midst of this, what are the chances that the message will be transmitted correctly?

The entire environment around you is the Channel in this example - as you can see hospital wards are not ideal channels.

However many other ways of communicating have equal potential for distortion of the message:
Fax or photocopy - poor resolution
Mobile phone - poor coverage/reception
Internet - poor bandwidth (= video stops streaming, or is poor quality)

Of course much depends on how the Sender constructs the message - but the point is that even if it is very clearly done, distortion can still ruin the communication process.

Solutions, not problems

Ok so how can we avoid this problem as best we can?

Step 1 - Good message structure
It has been shown that structuring data in different ways can change the way a clinician perceives it.
Elting et al presented data from a trial in 4 different ways - Table, Pie Chart, Bar graph and Icon - and asked clinicians to interpret the data. Depending on the way it was presented, the response varied.
BMJ. 1999 Jun 5;318(7197):1527-31. Influence of data display formats on physician investigators' decisions to stop clinical trials: prospective trial with repeated measures. Elting LS, Martin CG, Cantor SB, Rubenstein EB.

Correct interpretations were more common with Icon (82%) and Table (68%) that with pie charts or bar graphs (both 56%).

The data was exactly the same, it was only the way the message was structured that changed. This is obvious when you think about it - if you were asking someone out, the way you pose the question has serious repercussions on your chances of success...

Step 2 - Be aware of, and avoid distortion
Now you know distortion exists everywhere, you should try to minimise its effect. Choose methods of communication that are clear and simple. Make sure you do this in a suitable environment. For example, a doctor telling a family their son is very sick, in a busy hospital corridor, is not only rude and inconsiderate - it also decreases the chances of the message getting across correctly.

Step 3 - Check the message has been received correctly
You will be told several times to check with patients that they understand what you are saying - this is why! By asking them to repeat it, you can check what message has been received, and clarify any misinterpretations.

Here's a good reason:
In many countries, the main reason that patients complain about healthcare staff is poor communication. If you get it right, not only will it save you time and reduce clinical error, it will also make you less likely to be sued or get into trouble...

Grice's Conversational Maxims

What do you think of this idea:

The philosopher Paul Grice proposed a set of four maxims, which lay out a set of rules about conversation. Read them carefully and ask yourself if you break any of the rules!

1. Maxim of quantity: Say only what is needed.
   Be sufficiently informative for the current purposes of the exchange
   Do not be more informative than is required

2. Maxim of quality: Make your contribution one that is true.
   Do not say what you believe to be false
   Do not say that which you lack adequate evidence for

3. Maxim of relevance: Say only what is pertinent to the context of the conversation at the moment

4. Maxim of manner:
   Avoid obscurity of expression
   Avoid ambiguity
   Be brief
   Be orderly

Clearly you are not going to be a very interesting person to talk to if you employ this strategy in all your conversations, but do you think this could be a helpful tool in professional situations? Or is it a set of ridiculous limitations?

Self-reflection: Poor communication / Grice's maxims

Think of an example of a situation where you saw poor communication.

1. What were the problems? What was the result?
2. If you apply Grice's maxims to the situation does this remedy it?

Communication Systems

We've already discussed Systems in general, now let's look at this specific example.

Clearly, effective communication is essential to the running of a good health service. Some hospitals have lots of old information systems, that cannot communicate with each other. But is a "Communication System" only made up of computers?

Coiera argues that it is composed of people, messages, mediating technologies and organisational structures. The requirement for people and messages seems intuitive. Mediating technologies and organisational structures may not seem to be. But an example would be the Electronic Health Record (EHR).

In hospitals that have an EHR you might think that most exchange of healthcare information is done electronically. In fact one study in Emergency Departments showed that it only represented 10%. So the other 90% of information transactions was directly between staff. The scale and complexity of information transfer means that errors are highly likely. In a retrospective review of 14,000 deaths in hospital (Wilson et al, 1995), communication errors were found to be the lead cause (actually twice as frequent as inadequate clinical skill).
Med J Aust. 1995 Nov 6;163(9):458-71. The Quality in Australian Health Care Study. Wilson RM, Runciman WB, Gibberd RW, Harrison BT, Newby L, Hamilton JD.

Another study (Bhasale et al, 1998) showed that around 50% of all adverse events in primary care were associated with communication difficulties.
Med J Aust. 1998 Jul 20;169(2):73-6. Analysing potential harm in Australian general practice: an incident-monitoring study. Bhasale AL, Miller GC, Reid SE, Britt HC.

There is clearly a very considerable benefit to be gained from improving communication, and this can be measured in terms of morbidity and mortality.

Here is a recent example:

A 6 year old boy was assessed by a surgeon, who decided to remove his diseased right kidney. Unfortunately he wrote the wrong side in his notes.

On admission to hospital, the mistake was not noticed. The child was booked for left kidney removal. When his parents suspected the wrong operation was being considered, they communicated their concern to a nurse and a junior doctor. The father said he became frustrated as he could not get a definitive answer as to which side was to be operated on.

The surgeon then asked a junior doctor to perform the operation. This doctor consulted the notes, but not the imaging. He removed the wrong kidney. Despite the parents alerting staff, the error still occurred.

This was a preventable and catastrophic error. Do you think poor communication was a major cause?

Handover

This is the process whereby staff handover care to other staff, usually at change of shift. It is an obviously risky procedure as communication can fail and one mistake can be enough to start a chain reaction of adverse events.

Although nurses have been doing this for years and years (many at 8am and 8pm every day, when the shifts change), doctors and other staff haven't been as conscientious. In one study, 35% of Australian doctors reported that there were no set procedures for handover where they worked.

Aust Health Rev. 2005 Feb;29(1):68-79. A description of handover processes in an Australian public hospital. Bomba DT, Prakash R.

In another, 83% of UK doctors believed that the handover process they used was poor.
J R Coll Physicians Lond. 1996 May-Jun;30(3):213-4. The junior doctor handover: current practices and future expectations. Roughton VJ, Severs MP.

Here is a piece of news coverage that gives an example of how handover can go wrong:
Communication at handover (Australian Medical Association)

And a possible solution - view the OSSIE Guide to Clinical Handover Improvement
Read the following:

Clinical Handover - some facts, on page 3
Then skim through Chapter 1 (p5-10) to get an idea of how one proposed structure for the process of handover looks.

Can computers help?
So clearly getting communication right is extremely important. Get it wrong, and you can injure or kill patients and thus be sued.

Is health informatics going to make us better communicators?

Well, clearly some steps towards change have been taken in recent years. There is more emphasis on communication skills in education. Patient safety advocates have been showing that the type and content of communication needs to change to make healthcare safer. We have been advised to take a leaf out of the book of airline pilots, who have an open communication policy, and a questioning environment so that mistakes should be avoided.

As well as this effort to promote good communication and make it a core skill, there have been lots of attempts to get computers to help out. But the initial research was worrying. For example it was shown that the presence of a computer during consultations had several bad effects. Doctors with computers in the consulting rooms confined themselves to short responses, delayed responding, glanced at the screen in preference to the patient, or structured the interview around the computer rather than the patient. They also gave the patient less information.
Fam Pract. 2010 Dec;27(6):644-51. Epub 2010 Jul 26. Consulting room computers and their effect on general practitioner-patient communication. Noordman J, Verhaak P, van Beljouw I, van Dulmen S.
So clearly we can't just put technology into a situation that already works reasonably well, and not expect it to change things. This socio-technical change is very important to consider and we will explore it more shortly.

Summary
Communication can be distorted by many things in healthcare – the message received may not match the message sent

Grice's Maxims can help to structure messages better
Structured systems of handover should reduce clinical errors
Technology can affect how we communicate with our patients

Chapter 9: Decision Support Systems

Do you trust your doctor? How do you know they are doing the right thing - do you google what they tell you when you get home?

Once upon a time, doctors were thought to be experts in all areas. Anything they said was respected as the truth. If something didn't work, there was always another explanation.
And so the old saying developed, "Doctors differ and patients die".

But why should this be? Questions began to be asked as to why one doctor would treat the same disease differently to their colleagues. Was it inspired genius or simply a lack of knowledge of the appropriate action? This might not be clear until the outcome was known – a good outcome and they did the right thing; a bad outcome and they should have done what other doctors do.

The same applied to new treatments. For example, the clot-busting agent Streptokinase was shown to be useful in heart attack in 1958. By the early 1970s there was convincing evidence of its efficacy. In the early 1980s a meta-analysis showing its value was published. However it took until the late 1980s before it was formally recommended as treatment to be considered by all doctors.

This gap between establishing efficacy and this information filtering down to clinical practice has been shown in several examples. Conversely, there is often a delay when a drug is withdrawn from the market (for example because it was found to be unsafe), before doctors become aware of its demise. Doctors might still be prescribing it months later, even though it's not available. Patients may be even farther down the information chain.

Protocols
Nowadays everyone in healthcare recognises that evidence-based practice is the only way to go. Why would you treat someone in a particular way, unless you knew that there was evidence of its effectiveness? So, there has been a move towards standardising treatments, using "protocols". A protocol is a set of instructions, based on good clinical practice. Clearly a protocol is only as good as the evidence it is based on, and this should be clearly identified. But the idea is that increased compliance with protocols by doctors and other healthcare practitioners should result in less varied care and fewer poor outcomes...

As with any other intervention, we would of course want to see evidence that protocols are effective. In one review of 59 protocols all but four were found to improve the process of care. However protocols are not always followed and indeed some have actually been harmful – usually because they were poorly thought out or are missing vital information.

Guidelines
In recognition of the fact that clinicians are not simply robots, the term "Guideline" has become more commonly used. This takes into account the fact that guidelines do not cover all clinical scenarios, and indeed in some scenarios they may not be

appropriate. The current position is that clinical guidelines are there to assist clinicians - and also patients (a guideline should not allow a doctor to ignore what the patient's wishes are). However if a doctor departs from an established guideline there should be a clear reason why. The Courts have even expressed an interest in this idea in clinical negligence cases, and it would be difficult to justify not employing a guideline if a poor outcome resulted. Nevertheless guidelines are not supposed to constrain us, nor force us to act in an appropriate way.

Guidelines must be:
Up to date
Clear
Easy to follow
Useful
Applicable to local circumstances
Evidence based

How to make a guideline
Many of you will be involved in writing guidelines at some stage when you are in clinical practice. So how is it done?

First why not have a look at one, to see what we are talking about.

Task: Find a guideline

Search on the internet for a clinical guideline in an area that is of interest to you. Note the following:
1. The name of the guideline
2. Which of the ideal characteristics above does it have?
3. How could you improve it?

Guideline development is initiated when a clinical question is posed. For example, how should I treat a cruciate ligament strain?
The next step is to form a development group, decide on the scope of the guideline and start a literature search (good thing you already learned how to do this!) It doesn't stop there though - once you have found evidence you have to appraise it. Only good evidence will do! Once there is a consensus you can start to form recommendations.

As we said, a guideline is no good if it's not used - so you need to regularly audit this.

What's an audit?
Audit is when clinicians look at how good they are at following guidelines. So for example in my practice I regularly see stroke patients. Every few months we look at our management of these patients and see how well we are compared to the guidelines. The type of thing we look at is: Have they had a CT scan within 24 hours? Have they had a physiotherapy assessment within 2 days? Are they checked for depression?...

When developing a guideline you agree on audit criteria. Then you write the guideline document, send it to stakeholders (a common healthcare management term, that simply means "everyone who might be interested") for review and get external appraisal. Finally you can disseminate (i.e. publish and spread) it...

Of course it may be out of date by then - the above process can take 2 years to accomplish. So regular review is necessary.

Here is an excerpt from a guideline commonly used by doctors, from the British Thoracic Society Guidelines for the Management of Pneumothorax (collapsed lung)

• Needle aspiration (NA) is as effective as large bore chest drains, and may be associated with reduced hospitalisation and length of stay (A).
• NA should not be repeated, unless there were technical difficulties (B).
• Following failed NA, small bore (<14Fr) chest drain insertion is recommended (A).
• Large bore chest drains are not recommended for pneumothorax (D).

The letters after each statement refer to the grading of recommendations - see website below for more detail on this:
Patient.co.uk - Levels of evidence

NICE (the National Institute for Health and Clinical Excellence) in the UK is an independent organisation responsible for providing national guidance on the promotion of good health and the prevention and treatment of ill health. They make guidelines on lots of things and they are used in many other countries too.

Task: How to make a guideline - NICE

Go to the NICE website below and skim-read the Introduction to the manual. In particular look at the flow chart of the guideline development process.

NICE Guidelines Manual

Take the opportunity to have a quick look at the NICE website - you can search to see what guidelines they have in their database.

Computer-based protocol systems
In many hospitals a Clinical Guideline is posted on the noticeboard, or filed in a folder, or stuck onto a cupboard. This doesn't seem to be the best way to disseminate them, or to get clinicians to remember to use them. So can computers help here?

There are two types of computer-based protocol systems:

1. Passive protocol systems

this delivers information only, and does not directly affect a clinician's treatment options
Essentially this involves simply putting the written sheet on a computer screen instead (not literally, although I have seen this done).

One trial of this type of system showed that simply having guidelines on a computer improved adherence by 19%. However another showed no difference.

2. Active protocol systems

This is a more interactive method, in which a clinician's actions are guided by the system's knowledge
A systematic review of active protocol systems showed that in 14 of 18 systems, guideline adherence improved.

Clinical Decision Support Systems
Increasingly, the active system is being used as the basis for clinical decision support systems (CDSS). This means that as a physician uses software to make clinical notes, or prescribe medications, the system is monitoring the activity and triggers suggestions or reminders.

One of the commonest examples is computerised prescribing. These systems have actually been introduced into clinical practice, and have reduced medication errors. If you are trying to prescribe a medication the system helps you to choose an appropriate dose and administration method, checks for interactions with other medications, verifies it's in stock (and still on the market...) and importantly won't let you prescribe the wrong dose.

A CDSS does this by using knowledge of medications, prescribing rules and clinical guidelines. For example it may ask you why you are prescribing moxifloxacin when co-amoxyclav is the local recommended antibiotic for community acquired pneumonia. There is also an argument for it telling the physician how much his prescriptions are costing the hospital or patient. This could be easily done.

There are other cases where it could be helpful: e.g. by linking with laboratory data it could automatically stop administration of a drug if the level in the bloodstream is too high, or if the patient is in kidney failure...

This sounds like it has real potential to improve patient care.

Pros and cons of decision support

Advantages of effective CDDS
Can automatically complete entries in the clinical notes
Recommendations can be based on a situation or an alert
Data display can be modified and adapted as required
Increased quality of decisions, possibly improved outcomes

Increased consistency of care
Educational effect on clinicians
Less requirement to read literature to stay up to date

Disadvantages of CDSS
Expense
Can take longer to do same task
Needs training
May be very complicated to use
Poor clinician uptake
Clinicians can ignore the advice (this happens especially if the system "floods" you with alerts

Task: Communication and Clinical Guidelines
"Human Factors"

This video gives a good example of how horrific things can happen when communication stops, and guidelines are ignored.
(14 mins).

Some comments:

It is worth noting that in this example, nurses were aware of what should be done. But none felt they occupied an appropriate place in the clinical hierarchy to offer this opinion.
If the correct clinical guideline had been followed, there is a strong chance that a poor outcome would have been avoided.

Summary:
It is impossible to stay up to date with all relevant information so we need to condense clinically relevant evidence into guidelines.
These can then be used in electronic decision support systems to help ensure we implement the evidence in our daily practice.

Chapter 10: Human-Computer Interaction

This concept, which sounds a bit science fictiony, is easy to understand but actually very complex in its workings.

The basic idea is that *the way we use* computers, or software, has a huge influence on how well it works. The discipline of Human Factors explains how the interaction between humans and technical systems is of particular interest in the area of patient safety.

Think back to the example I gave you in Chapter 1 - the Therac-25 Linear Accelerator Disaster. Patients were given 100 times the intended dose of radiation. The machine gave error codes, but since the staff did not understand them, they overrode them and forced the machine to continue dispensing radiation. The humans were at fault here.

On the other hand, you will all have had experiences of where a computer just won't do what you want it to, whether it's deleting a document by accident, unhelpfully autocorrecting your spelling to American English, or my own example - sites that force me to leave out the apostrophe in O'Hanlon...

Usability - when computers don't help

The ultimate acceptance or rejection of any healthcare information system is whether it is used by the target population. The same applies to all technology. The easier it is to use, the more likely it is to do well. Why are iPhones uniquitous now, when there were only Windows Mobile smartphones a few years ago? They are simple to use.

Preece defines usability as "the capacity of a system to allow users to carry out their tasks safely, effectively, efficiently, and enjoyably". Does the iPhone come out tops here?

A system's usability is the key barrier to its acceptance. Doctors are notorious for wanting computers to work well for them - but in fairness most of the software we have had to use for the last few years has been terrible. In Sweden the introduction of an electronic health record system actually increased staff workload by 25%. This is a complete disaster as nobody wants to have to do more work. Another example is the US Dept of Defence system "Composite Health Care System II" which took 2 years and a million dollars a day to develop. It failed user acceptance testing. One of the main reasons was that users were not involved in its design and testing. So they didn't like the finished product.

If you want to know more about Usability, have a look at this book:
The Design of Everyday Things by Donald Norman - which gives simple rules to improve the usability of objects as diverse as cars, computers, doors, and telephones.

So - how do you know if a system is user-friendly? Is it easy to tell?

Self-reflection: Usability at work

Think of a piece of software, or a workstation, or a console or any piece of technology that you have seen.

1) What works well?
2) What doesn't work well?

Usability testing

A lot of work has now been done in this area and there is a framework for performing usability assessments. Which means there is no excuse for a computer system being difficult to use anymore!

It is also recognised that usability testing has to continue through several "iterations" - i.e. through initial development, initial testing, roll-out, even after the end-users start using software - the development needs to keep going.

We will look at some of the established methods of usability testing, and take as a case study the evaluation of a Decision Support System.

Why this is important for you

You should expect to become involved in usability testing in the next few years. As more new information systems, EHRs and Decision Support Systems come onstream, your feedback on how well they are working will be vital. If you know how usability testing is done you will be very valuable. Not to mention the fact that we will all get superior quality systems to use. The end result will hopefully be better, easier clinical practice and of course improved patient care.

How it works

Kushniruk states that usability testing is "a practical yet scientific approach to evaluating how usable our systems are and can also provide invaluable feedback to designers with ways of improving their usability, safety and workflow."

Essentially it involves observing representative end-users of a system actually using it, and collecting data about their interactions. They are asked to perform several representative everyday tasks and are video-recorded while doing so. They talk through their experience and report their problems. Kushniruk believes that by using this technique, "the majority of usability problems with a system can be identified and recommendations made for fixing them within a short period of time".

The methods borrow from several disciplines: cognitive psychology, computer science, systems engineering, and usability engineering. They represent a shift in focus on the design of software and systems to actually understanding how end-users interact with them.

Kushniruk describes the approaches to usability testing as follows:

(a) characterizing how easily a user can carry out a task using the system,
(b) assessing how users attain mastery in using the system,
(c) assessing the effects of systems on work practices, and
(d) identifying problems users have in interacting with systems.

Conventional approaches such as a survey as to what users think of a new piece of software are very limited. With proper usability testing, you can observe the user as they interact with the system. You can ask them to perform certain tasks to see if it is intuitive. It's also the only way to see if people try to do things the system shouldn't let them, or go about things in ways the designer never predicted.

For example - some EHRs prompt the clinician to ask questions of the patient. Changing the order of the questions can have a dramatic effect on the interview. Imagine if someone designed a system to ask Social History before History of Presenting Complaint... Such things can have a profound effect on the way we interact with our patients, and the way our thought processes work. It's important that clinicians are involved so that the design makes us work well, rather than introducing new barriers or potential sources of error.

Methods of usability testing
Heuristic evaluation: judging the user interface and system functionality to see if they conform to established principles of usability and good design (called heuristics).

Guideline reviews checking for conformance with usability guidelines.

Pluralistic Walkthroughs: review meetings where users, developers and analysts go through specific scenarios together and discuss usability issues that might arise.

Standard Inspections: inspecting the interface to ensure compliance with usability or system standards.

Cognitive Walkthrough: a method which applies principles from cognitive psychology to simulate the cognitive processes and user actions needed to carry out specific tasks using a computer system.

Task: Usability testing

Skim-read the following article and answer the questions below:

Usability testing in medical informatics: cognitive approaches to evaluation of information systems and user interfaces.
A. W. Kushniruk, V. L. Patel, and J. J. Cimino
Proc AMIA Annu Fall Symp. 1997: 218-222.

Questions:
1. How do the authors define usability testing?

2. What is the "think aloud" method?

3. Who participates in the studies? (see Step 5)

4. Is this technique best used for formative or summative evaluation? (see "Experiences to Date")

Answers are below

.
.
.
.
.

Answers:

1. Usability testing refers to the evaluation of information systems that involves participants (i.e. subjects) who are representative of the target user population.

2. Subjects are asked to "think aloud", or verbalise their thoughts, as they perform specific tasks (e.g. entering data into a CPR system

3. In the authors' studies participants include the subject (e.g. health care worker), a test administrator whose responsibility is to ensure that the session proceeds properly, and in the case of studies involving doctor-patient interaction, either a real patient, or a "simulated patient" (i.e. a research collaborator, who "plays" the part of a patient).

4. Cognitively-based usability testing can be applied throughout the life cycle of information systems, i.e. from testing and evaluation of early prototypes (i.e. formative evaluation) to final, or summative, evaluation to determine if a computer system has met usability criteria. The authors believe the greatest benefits come from formative analysis, where there is an opportunity for results to be communicated to the designers of the information system and appropriate improvements to a system can be made on the basis of these results.

What kind of results do you get from this?

One usability study looking at doctors using a Clinical Decision Support System noted:

Two physicians ignored the antibiotic advice the system gave (for antibiotics by mouth), and prescribed intravenous levofloxacin instead. (Is this a compliance issues with physicians' use of guidelines?)
There were 2 potentially life threatening mistakes (by users) – both involved failure to prescribe antibiotics. (if you are used to doing this on paper it can take time to become familiar with it as a step on an electronic system)

To conclude:
It's important to remember that any new intervention in health care can potentially cause harm. This quote highlights the danger of not performing comprehensive usability testing:

"it would be unthinkable that the airline industry would have its first trial of an airplane's flight capabilities with real passengers, but something not far off is currently occurring in the healthcare field, where systems are being rapidly implemented with no true understanding of their risks and benefits."

Summary:
When designing information systems it is essential to consider how humans and computers interact.
Socio-technical change may occur when introducing new systems.
Usability testing can help to predict how systems will affect behavior.
Usability is extremely important – if a new system takes longer, users are less likely to embrace it.

Endword

Health informatics is something we all use and are familiar with – we just don't realize it until we start to look more closely at it.

There is no doubt that informatics is becoming more relevant and as electronic health records, decision support systems and e-health interventions such as telemedicine become more advanced we will find ourselves interacting with it on a daily basis.

Become aware of some of the key points in informatics is therefore vital for us to do our jobs properly as clinicians.

I hope that this short book has provided a meaningful introduction for you and has opened your eyes to an exciting area of work.

Finally it is worth noting that in these challenging times, it is likely that a large informatics-trained workforce will be required. If you like the area, get yourself a qualification and you will be even more useful!

Enjoy the trip…